HENRY
THE FOURTH

by Stuart J. Murphy

illustrated by Scott Nash

HARPERCOLLINSPUBLISHERS

LEVEL 1

To Nancy the First!
—S.J.M.

For four old friends—Bandit, Black Dog,
Clovis, and Maggie—good dogs all
—S.N.

HarperCollins®, ✦®, and MathStart® are registered trademarks of HarperCollins Publishers.
For more information about the MathStart series, write to
HarperCollins Children's Books, 10 East 53rd Street, New York, NY 10022, or
visit our web site at http://www.harperchildrens.com.

Bugs incorporated in the MathStart series design were painted by Jon Buller.

Henry the Fourth
Text copyright © 1999 by Stuart J. Murphy
Illustrations copyright © 1999 by Scott Nash
Printed in the U.S.A. All rights reserved.

Library of Congress Cataloging-in-Publication Data
Murphy, Stuart J., date
    Henry the fourth / by Stuart J. Murphy ; illustrated by Scott Nash.
      p.   cm. — (MathStart)
    "Level 1, ordinals."
    Summary: A simple story about four dogs at a dog show introduces the ordinal numbers:
first, second, third, and fourth.
      ISBN 0-06-027610-X. — ISBN0-06-027611-8 (lib. bdg.) — 0-06-446719-8(pbk.)
      1. Numbers, Ordinal—Juvenile literature.   [1. Numbers, Ordinal.]   I. Nash, Scott, date, ill.
II. Title.   III. Series.
QA141.3.M87   1999   98-4960
513—dc21   CIP
                                                                                       AC

        Typography by Al Cetta   1 2 3 4 5 6 7 8 9 10   ✦   First Edition

"Welcome to the dog show!"
shouted Jeremy.

"Take your places," he called to the dogs.

"Today, you will see tricks
performed by the most talented
dogs on the block,"
announced Jeremy.
"Maxie, you're first."

1st   **2**nd   **3**rd   **4**th

Joel said, "Speak, Maxie. Speak."
Maxie barked loudly.

"Baxter, you're second," called Jeremy.

1st

2nd

3rd

4th

Suzi said, "Beg, Baxter. Beg."

Baxter begged beautifully.

15

Jeremy hollered,
"Daisy will be the third dog."

Jennifer said, "Roll over, Daisy. Roll over."

Daisy rolled over and over.

19

"Henry, you're fourth," Jeremy said.
"We're waiting."

1st **2**nd **3**rd **4**th

Tom pulled and tugged.
But Henry wouldn't budge.
Then Jeremy took out a dog treat.

Suddenly, Henry bounded toward Jeremy and tackled him with a big sloppy kiss!

Everyone laughed and cheered.

Then Jeremy announced:

"First, Maxie
spoke loudly.

Second, Baxter
begged
beautifully.

Third, Daisy
rolled over
and over.

"But Henry the fourth is the king of the show!"

I f you would like to have more fun with the math concepts presented in *Henry the Fourth*, here are a few suggestions:

• Read the story with the child and talk about what is going on in each picture. Discuss the fact that the dogs are in a line and that they appear onstage one after the other.

• Ask the child to point to the dog highlighted in any math diagram. Talk about the position of that dog in relation to the other dogs. Ask questions such as "How many dogs came before Baxter?" and "Where is Baxter in the line of dogs?"

• Encourage the child to tell you the story. Point out the relationships between numbers and ordinals. For instance, Daisy is number three. She is the third dog in the show.

• Practice writing the name of each dog and the ordinal that describes its position in the story. Maxie: 1st, Baxter: 2nd, etc.

• Talk about things around the house and identify their order. "Which step is first going up the stairs? Which is second? Which is third?" Look at cars parked on the street and talk about their order. "Which is the second car from the corner?"

• Use the language of the story in everyday situations. Talk about the carpool. "Who gets picked up first?" "Who is third?" Discuss favorite TV programs. "Which program comes on first?" At the supermarket, who is first in the checkout line? In which position are you?

Following are some activities that will help you extend the concepts presented in *Henry the Fourth* into a child's everyday life.

In the Playroom: Place 5 to 10 crayons of different colors in a row. Encourage the child to identify which color is first, second, etc. Ask questions such as "In which place is the green crayon?" "How about the yellow crayon?" "Can you name the order of the crayons backward?"

Parade: Make a line of toy animals or cars. Pretend it's a parade. Ask the child to point to the second one in line, then the fourth, the third, and the first. Turn them around. "Now which is first?"

Showtime: Encourage a group of friends to put on a talent show. Each friend can perform a different act or trick. Let them decide who will be first, second, third, etc. Ask them to announce their acts by the order they're in: "First we have . . . " "Our second act is . . . " "Fourth and last . . . "

The following stories include similar concepts to those that are presented in *Henry the Fourth*:

- PANCAKES FOR BREAKFAST by Tomie DePaola
- FIRST, SECOND by Daniil Kharms
- NOEL THE FIRST by Kate McMullan